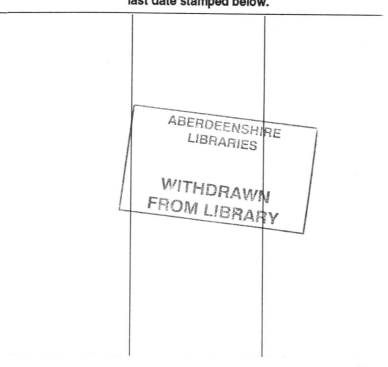

Islands

Catherine Chambers

Heinemann
LIBRARY

First published in Great Britain by Heinemann Library,
Halley Court, Jordan Hill, Oxford OX2 8EJ,
a division of Reed Educational and Professional Publishing Ltd.
Heinemann is a registered trademark of Reed Educational & Professional Publishing Limited.

OXFORD MELBOURNE AUCKLAND
JOHANNESBURG BLANTYRE GABORONE
IBADAN PORTSMOUTH NH (USA) CHICAGO

Designed by David Oakley
Illustrations by Tokay Interactive Ltd
Originated by Dot Gradations
Printed in Hong Kong/China

04 03 02 01 00
10 9 8 7 6 5 4 3 2 1

ISBN 0 431 09840 9

British Library Cataloguing in Publication Data

Chambers, Catherine
    Islands. – (Mapping earthforms)
    1. Island ecology – Juvenile literature 2. Islands – Maps –
    Juvenile literature
    I. Title
    577.5'2

Acknowledgements
The Publishers would like to thank the following for permission to reproduce photographs: Bruce Coleman
Limited: J Cancalosi p19, K Wothe p18; Corbis: p21; Empics: T Marshall p14; Robert Harding Picture Library: p5,
F Hall p20, R Richardson p11, P van Riel p23, N Wheeler p10, G Williams p13, L Wilson p27; Oxford Scientific
Films: H Bardarson p29, G Soury p26; Science Photo Library: US Geological Survey p4; Still Pictures: B Brecelj
p17, M Edwards p24, E Hussenet p6, H Klein p16, G and M Moss p8.

Cover photograph reproduced with permission of Robert Harding Picture Library.

Every effort has been made to contact copyright holders of any material reproduced in this book. Any
omissions will be rectified in subsequent printings if notice is given to the Publisher.

For more information about Heinemann Library books, or to order, please phone ++44 (0)1865 888066, or send
a fax to ++44 (0)1865 314091. You can visit our website at www.heinemann.co.uk.

Any words appearing in the text in bold, **like this**, are explained in the Glossary.

Contents

What is an island?

An island is a land mass that is completely surrounded by water. Some islands are tiny – you could walk all the way around them quite easily. Others are hundreds of kilometres long. Most islands lie in the sea near the coasts of large land masses, called **continents**. But some rise in the middle of the ocean. You can also find islands in rivers, lakes and in **deltas**, which is where rivers meet the sea. We shall be looking at different types and sizes of island all around the world.

How are islands formed?

Some islands are the peaks of volcanoes that rise from the ocean floor. Others formed when the sea level rose over mountains, leaving peaks sticking out of the water. A wide river sometimes carves its course

A group of islands is called an **archipelago**. The islands on the archipelago are usually formed in the same way. But the plants and flowers on each island in the archipelago can be quite different. Some archipelagos make up one country. The islands in this archipelago make up Hawaii, one of the states of the USA.

around an island of hard rock. Hard rock can also form an island in a lake. A slow-moving river splits up near the sea and winds around islands of fine soil.

When you think of an island you probably have something like this in mind. But many islands are covered in a huge cap or sheet of ice.

What do islands look like?

Small island landscapes range from hot **coral** sands to cold, windswept rocks. Large islands can have a variety of landscapes, from mountains and **plateaux** to flat river plains. We shall discover how the size of an island and its **climate** affects its landscape.

Life on an island

Islands are very **isolated**. So special **species** of plant and animal have **evolved** separately on them. But what about humans? Today, most island communities have constant contact with large continents. They are able to buy many goods that they cannot produce on their own. We shall also see how some island peoples still use the natural materials around them to build their homes and make a living. We will find out what the future holds for life on the islands of the world.

Islands of the world

Where in the world?

There are thousands of islands in many parts of the world. If you look at the map you will see that most of them lie near the **continents**. Some lie further out in the oceans and seas. You cannot see any islands in lakes, rivers or **deltas** on the map because they are too small. But most of these form in the largest lakes, the widest rivers and the biggest deltas.

It is sometimes difficult to decide what is an island and what is a continent. Greenland is called an island. It is the largest in the world – just over 2 million square kilometres (about 840,000 square miles). But Australia is called a continent, even though it is surrounded by water. It is thought to be too big to be an island. Australia is so large that it has several different **climates** and types of **vegetation**. Most big islands also have a range of temperatures and rainfall in different parts.

Baffin Island is the fifth-largest island on Earth. It has two huge lakes on it. The island belongs to Canada, which has more islands than any other country. Most lie off Canada's long coastline. But some have formed in the middle of the huge Great Lakes, which is the biggest system of lakes on Earth. The Arctic **archipelago** in the north includes 31 islands with an area of more than 1300 sq km. Canada's northernmost tip is marked by Ellesmere Island.

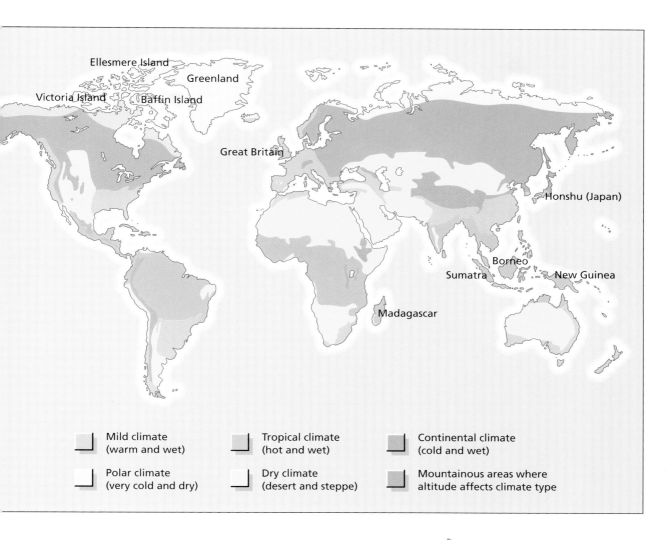

Mild climate
(warm and wet)

Tropical climate
(hot and wet)

Continental climate
(cold and wet)

Polar climate
(very cold and dry)

Dry climate
(desert and steppe)

Mountainous areas where
altitude affects climate type

Islands in different climates

Many islands close to continents lie in the cold northern part of the world, from Canada to Russia. Most of these islands were formed millions of years ago, after a time when the region was even colder – the **Ice Age**.

Small **coral** creatures live in great numbers in the warmest seas. Here, you will find the world's coral islands, which are made up of dead and living coral. Volcanic islands do not lie only where the climate is hot. They are formed where the Earth's crust under the ocean is weak – and very hot indeed!

You can see that islands are scattered all over the world. There are many in the middle of the Pacific Ocean although some are very small and are not marked on this map. But there are fewer in the middle of the Atlantic Ocean. A lot of the Pacific Ocean islands were fromed by volcanoes shooting up from the ocean floor.

7

How are islands formed?

Islands in the sea

Some islands were once mountains on the edges of **continents**. But during the last **Ice Age** they were covered in snow and ice. The **climate** began to warm up again about 10,000 years ago: the ice and snow melted and the sea rose. The waters covered the mountains on the coasts, leaving just the peaks sticking out. Norway has 3000 islands like this around its shores!

Other islands were formed when the sea filled a dip between a small piece of land and a continent. The dip was a large strip of land which slipped down along huge cracks in the Earth's crust. The British Isles were separated from Europe in this way and Sicily was **isolated** from Italy

This island is shaped like a ring, with a lake called a lagoon in the middle. It was formed from a volcano. In warm waters, **coral** creatures attach themselves to the volcanic rocks. When they die they leave hard, crusty layers. The original mountain sinks, leaving behind a coral island ring called an **atoll**.

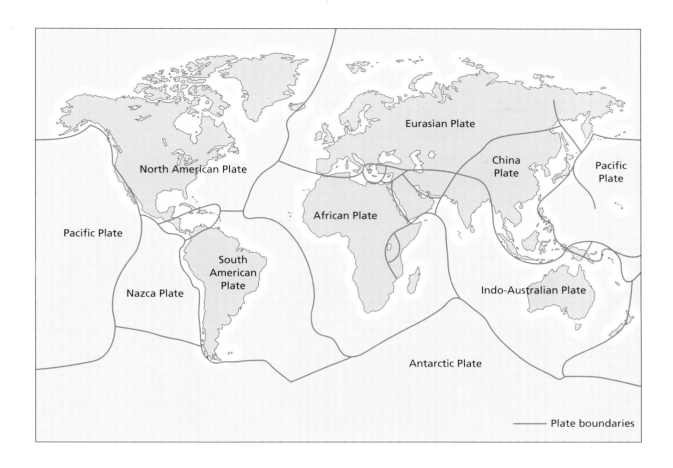

Some islands are formed of huge bars of sand, pushed up by ocean tides. Fraser Island lies off the east coast of Australia. It is 112 kilometres (70 miles) long and is covered in **turpentine** trees.

Lake and river islands

Some islands rise in the middle of wide rivers that wind their way along flat plains. The river **scours** out its course through soft rock and curves around either side of an island of harder rock. This happens in the middle of lakes too. Islands can also form in **deltas**. The river splits up in streams that have to make their way around islands of thick, muddy **silt**.

Hundreds of islands lie along the bold lines that you can see on the map. They mark the edges of **tectonic plates**. The plates sit on top of a layer of hot, sticky rock called the mantle. The islands form where the edges of the tectonic plates push together or pull apart, or they erupt suddenly through a weak point in the Earth's crust. The Azores in the Atlantic Ocean and Hawaii in the Pacific Ocean are volcanic islands.

Island landscapes

An island's landscape depends on how it was formed, its size and where it lies. Some big islands have a range of landscapes, while others look nearly the same all over. Some small islands rise as mountains with streams and waterfalls – or even hot, volcanic springs. Others can be bare rocks with no fresh water on them. An island's plant life plays an important part in the way its landscape develops.

Large island landscapes

Greenland is the world's largest island and is, in many ways, like other islands in the Arctic region. The main island is mostly hard, granite rocks which rise to a high, flat **plateau**. Eighty per cent of it is covered in a deep ice sheet, with just a few peaks sticking out at the top. Tightly curved inlets, called **fjords**, make a fringe all around Greenland's coast. The fjords were formed by **glaciers** carving their way from the upland down to the sea. Icebergs drift all around Greenland's shores. Many smaller islands surround the coast.

◈ Guam is a small island that lies in the Pacific Ocean. It has a **coral** island plateau in the north and volcanic hills and valleys to the south. Guam lies at the southern end of a submerged mountain range, which is 2500 km (1500 mi) long.

The island of Great Britain lies further south, where it is warmer. Great Britain was separated from the **continent** of Europe when a strip of land slipped, allowing water to flow over it. It has green hills, flat plains and just a few mountains. There are many rivers and lakes. A lot of the hills and lakes were worn by glaciers during the **Ice Age**. The glaciers also left behind mounds and plains of **eroded** rocks, sands and soils. Great Britain's coastline, though, has mostly been formed by waves battering the rocks.

Small island landscapes

Some small islands are volcanoes sticking out of the sea, like the Canary Islands which lie off the north-west coast of Africa. These have rich, fine volcanic soils on dry, mountainous slopes. Black volcanic sands surround the coastline. Some small islands in the Mediterranean Sea are not volcanic but they are stony, mountainous and very dry. In wetter parts of the world, streams and rivers also help to shape the landscape of small islands by eroding their course through the rock.

◈ This beach of black, volcanic sand is on Tenerife in the Canary Islands, a popular tourist destination. These islands often have rich volcanic soils in which grapes and other vegetables are grown.

Islands and the weather

The oceans and seas are where strong winds blow and where rain is made. Small, unprotected islands are easily affected. Larger islands often have different weather systems blowing on to them at the same time. It can be warm and sunny in one part of the island and cold and wet in another.

Wind and rain over the oceans

As wind blows over the oceans, the air picks up moisture from the sea. This moisture is held in the air as drops of **water vapour**, which rise and form clouds. The clouds shed rain when they are blown on to an island,

This map shows the main trade winds that blow between April and October. Sailors relied on the trade winds to blow their ships to trading ports. The trade winds often meet at islands in the **Tropics**, bringing heavy rain. **Monsoon winds** change direction for half of the year. In September and October they bring torrents of rain to the islands in the western Pacific Ocean.

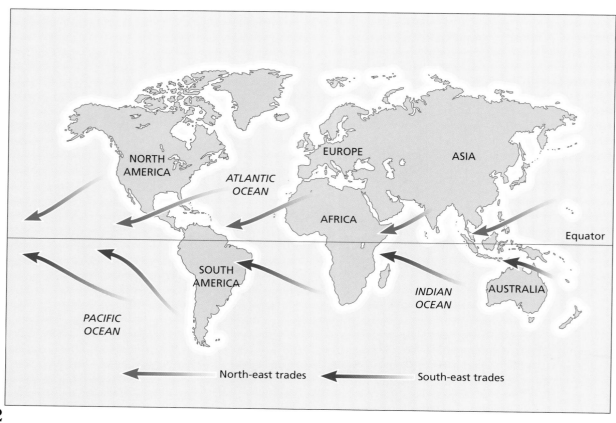

NORTH AMERICA

EUROPE

ASIA

ATLANTIC OCEAN

AFRICA

Equator

SOUTH AMERICA

INDIAN OCEAN

AUSTRALIA

PACIFIC OCEAN

North-east trades

South-east trades

especially if the island is mountainous. This is because the cloud rises up the mountain and the water vapour in the cloud cools and becomes water droplets. The clouds often shed all their rain on one side of the mountain, making the other side quite dry.

There are parts of the world where the winds blow over a huge stretch of ocean bringing a lot of rain to islands lying in their path. But the Greek Islands in the Mediterranean Sea are very dry. This is because the winds do not pick up a lot of moisture before they reach these islands.

Winds can also be affected by warm ocean **currents**, which heat up the air mass above them. The Hebrides off the north-west coast of Great Britain are warmed by the North Atlantic Drift. They lie as far north as very cold islands off the east coast of Canada, which have no warm current near them.

The islands of the Caribbean and the islands off the south coast of the USA are often badly hit by **hurricanes**. They occur when there are especially large amounts of very warm, moist air rising over the ocean. Hurricane winds can sometimes reach 250 km per hour (155 mi per hour).

Island life – Honshu

Honshu is Japan's largest island and the seventh-largest island in the world. On its west coast it is separated from China by the Sea of Japan. The Pacific Ocean stretches away from Honshu's eastern shores. Honshu was once a mountainous coastal strip on the edge of the Asian **continent**. But the land between the mountains and the continent slipped down and filled with sea water. Over 200 smaller islands lie around Honshu.

There are many volcanic mountains on Honshu, some of which are active. The highest is Mount Fuji, which rises to 3776 metres above sea level. Mount Fuji has very beautiful scenery and is a sacred peak for Japanese people. Many rivers begin in the mountains, which also contain lakes. Honshu has three of Japan's largest rivers – the Tone, Shinano and Kino.

Honshu has developed winter sporting facilites on its mountains. It hosted the 1998 Winter Olympics at Nagano in the northern part of the island. But tourism is not the most important industry on Honshu. Cars and electrical goods are some of the main industries in the island's many large cities, most of which lie on the mild east coast.

There are many rivers and lakes on Honshu because a lot of rain reaches the island, especially between September and October. This is when the **monsoon winds** bring storms and floods to the island from the south-west. They bring **hurricanes** too, which are known as typhoons here in the western Pacific.

Summers on Honshu are hot and humid, which means the air is moist. The temperature can reach 35°Celsius. But winter brings different types of weather to different parts of the island. The west coast is cold and the mountains are freezing. But the warm Kuro Siwo **Current** brings milder temperatures to the east side of Honshu. It is here on the lowlands that most people live. The different temperatures give a variety of **habitats** for many types of plant and animal.

Honshu lies in the 'Ring of Fire', where there is a lot of volcanic activity (see the map on page 9). Honshu has hot-water springs and **geysers**, as well as active volcanoes. There are also frequent earth tremors.

Island plants

Islands rise in many different **climates**. They are different sizes and have a variety of landscapes. So there are many types of island plant, as you can see from the pictures below and opposite. Island plants often have to cope with strong winds. Some islands are very dry and rocky, while others are hit by fierce rainstorms.

Isolated islands

Islands have some small plants that can be found almost anywhere in the world. These are **algae** and a few mosses and ferns. **Fungi** and **lichens** also grow freely. Islands that were formed when they were cut off from a **continent** often have unique **species** of larger plant. But many came originally from the continents nearby. Fine seeds can be lifted by the wind and blown

Arctic islands are frozen, dry, dark and very windy during the very long winters. Only tiny mosses and lichens cling to the small patches of bare rock. They grow thickly where they are not eaten by **caribou** and reindeer. When the short summer comes, only the top layer of soil thaws. Then, tough grasses and stubby, creeping shrubs grow. Flowering saxifrages have tightly packed flowers to protect them from the wind. Low-growing berry bushes hug the ground. This vegetation is called 'tundra', which means 'treeless plain'.

on to islands. Seeds are dispersed in bird droppings. Large seed capsules, such as coconuts, can float over the oceans. Nearly all the rainforest trees have been brought to islands by people.

Volcanic islands, especially, were originally totally bare. But most have been colonized, or taken over, by plants from other lands. The Hawaiian Islands are volcanic and have very lush **vegetation**. It is often said that 80 per cent of Hawaiian plants can be found nowhere else in the world. But in fact many of them developed from seeds carried by ocean **currents** and the wind. Fruit-eating **migratory** birds left seeds behind before they flew off again. But what is special about islands is their **isolation**. Over a long time, this has made some plant species develop into sub-species, which means they have features of their own.

The island of Jamaica lies in the Caribbean Sea, near Central America. Its climate is hot and moist and its landscape has low-lying plains and high mountains. So Jamaica has a great variety of plant species and thick, lush vegetation. This includes at least 2000 species of flowering plant. Many flowers are very big and bright, to attract the insects and birds that **pollinate** them. There are many varieties of palm and forests of precious hardwood trees, such as mahogany and rosewood.

17

Island creatures

Many **mammals, amphibians, reptiles**, fish, insects and birds live on islands with lots of **vegetation** and mild or hot **climates**. Islands in the **polar regions** are so cold that they have very little vegetation and many fewer varieties of animal. Here, most **species** live on the shores of the islands, which are thriving **habitats** for seals and seabirds, such as penguins.

Madagascar is a large island that lies in the Indian Ocean near the east coast of Africa. Like many islands, it does not have any really large mammals. But it does have several unique species of **primate**, called lemurs.

But the special thing about island creatures is that they are often unique species. This means that they have **evolved** differently from their relatives on nearby **continents**, or even on islands very close to them. Animals often have to compete for food, and competition on small islands is very fierce. So each of the species has evolved to eat a particular type of island vegetation. This makes sure that none of them starves. It also makes sure that one type of food does not get eaten so much that it dies out.

The Galapagos Islands are a group of volcanic islands in the Pacific Ocean. They lie about 1000 kilometres west of South America. Creatures on each of the islands are surprisingly different from one another. Giant tortoises have adapted to the contrasting island environments and even look different from each other. The tortoises from the dry islands have developed long necks and high curves in the shell where the neck stretches. This is so that the tortoise can reach up to high cactus branches and spiny cactus leaves. On wetter islands the tortoises have adapted to eating grasses and low-growing plants. These tortoises have shorter necks and a scooped bottom shell so that they can bend their heads down easily.

Tasmania is an island not far from the southern tip of Australia. It is home to the unique Tasmanian devil, a ferocious, meat-eating marsupial. A marsupial is a mammal with a pouch into which its young are born.

Island peoples

Since ancient times, communities have sailed from **continents** to islands in search of a better place to grow crops and raise animals. The long island coasts have provided plenty of fish, and harbours for boats. Some islands, especially those of volcanic origin, have deposits of gemstones and **minerals** such as gold. Today, small islands attract people from the mainland who want a simple, peaceful way of life. But young islanders often have to leave their homes to find work in bigger, busier areas.

Hong Kong island lies near the coast of mainland China. The island has no natural resources, but it is one of the busiest trading centres in the world. This is because it lies in the path of many trade routes and has deep, natural harbours around its coast.

Making a living

Frozen islands, whether big or small, do not have a large population. There is not enough unfrozen land to grow crops or raise animals. Minerals are very difficult to extract from the ground. But some Arctic islands do support communities of Inuit, who have managed to use their environment to survive. They fish and hunt **caribou**, reindeer, seals, whales and walruses.

Some islands are hot, rocky, mountainous and poor for farming, except around the coastal lowlands. The peoples of the Greek Islands have lived in these conditions for thousands of years. But they have made their living from fishing, shipping and trading. Now, these islands are popular with tourists, as are many of the hot islands around the world.

Volcanic islands have good soils, which have attracted farmers. However, volcanic islands can be dangerous places to live. The island of Montserrat, in the Caribbean Sea, erupted in 1996 and most of the inhabitants had to leave. Volcanic islands with hot springs and **geysers** can provide energy. In Reykjavik, the capital of the island of Iceland, the central heating is provided by hot water that is piped from geysers.

Ellis Island lies in New York harbour in the USA. When people from all over the world sailed to America to live, they stopped here first to have their documents checked. Islands have also been used to imprison and **isolate** people. Nelson Mandela was imprisoned on Robben Island off the coast of South Africa for 23 years.

A way of life – the Polynesians

Polynesia is a group of widely scattered islands in the central Pacific Ocean, from Hawaii in the north to New Zealand in the south. Over 2000 years ago, Polynesians began to leave their homeland, which was possibly in the Malay **Peninsula.** They travelled eastward on large wooden rafts to find new settlements. The peoples of Polynesia still use their natural environment to make a living and build a home.

The islands of Polynesia lie in a triangular area with its three points at New Zealand, the Hawaiian Islands and Easter Island. They are far from **continents** and are very open to natural disasters such as hurricanes and **tsunamis**.

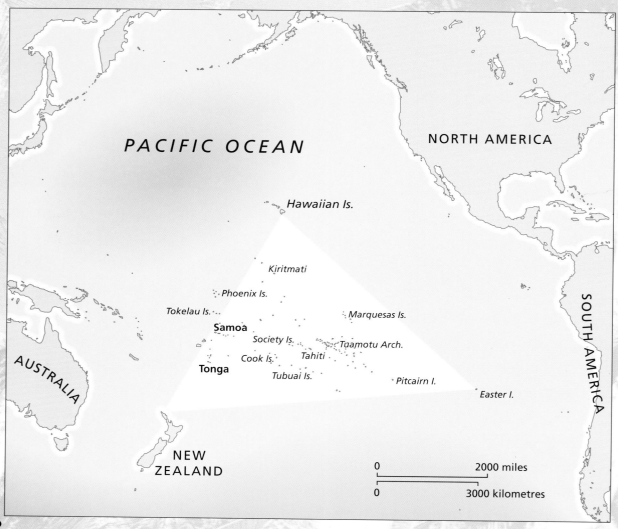

PACIFIC OCEAN

NORTH AMERICA

Hawaiian Is.

Kiritmati

Phoenix Is.

Tokelau Is.

Marquesas Is.

Samoa

Society Is.

Tuamotu Arch.

Cook Is.

Tahiti

Tonga

Tubuai Is.

Pitcairn I.

Easter I.

AUSTRALIA

SOUTH AMERICA

NEW ZEALAND

| 0 | 2000 miles |
| 0 | 3000 kilometres |

Building a home

Some of the Polynesian islands are volcanic, while others are made of limestone or **coral**. There is often lush **vegetation** with many trees, including palms and precious hardwoods. Traditional houses come in many styles. But most are made from a frame of hardwood posts with walls of bamboo and plaited palm leaves. The roof is thatched with reeds. Nowadays, many people live in modern houses that are built to cope with **hurricanes**.

The volcanic islands are especially **fertile** for growing crops. One of the main foods is taro. This is a root-crop which is baked, pounded and made into soft dough. Yams, vegetables and tropical fruits are also grown.

Coconut palms provide more than food. The fibres on the outer husk are called coir, which is made into rope and used to stuff furniture. Coir and leaves are woven into baskets and matting. Coconut oil is used in cooking and to make skin creams and soap.

Fishing is one of the main industries in the Polynesian Islands. Fishing canoes are carved out of solid pieces of wood taken from hardwood forests. They have beautifully carved **prows**. Modern industries include processing and packing the natural plants and fruits on the islands. Gold and other **minerals** are mined on some of the volcanic islands.

The islands of Polynesia are well known for their highly decorated carvings and cloth designs. Each island has its own style. Many tourists buy carvings and decorated cloth, which is made from tree bark and other natural fibres.

Our changing islands

Natural changes?

Islands change all the time. Some rise up from the ocean floor, or sink down under the waves. Others grow as **coral** deposits which build up layers of land. The sea constantly batters the coastline, wearing it away and changing its shape. Inland, strong winds and torrential rain **erode** mountain landscapes and widen river valleys.

In recent times, some of these changes have happened very quickly. **Hurricanes** are fiercer and more frequent. In some parts of the world, coastlines are eroding so rapidly that cliff-top villages are falling into the sea. Some islands are sinking fast. Both of these changes are occurring partly because there is more water in the oceans. The **ice caps, glaciers** and icebergs of the Arctic and Antarctic are melting, making sea levels rise.

Haiti lies on the Caribbean island of Hispaniola, which it shares with another country, the Dominican Republic. Haiti's lush forests have nearly disappeared, leaving the land quite bare. The Dominican Republic has managed to keep its trees.

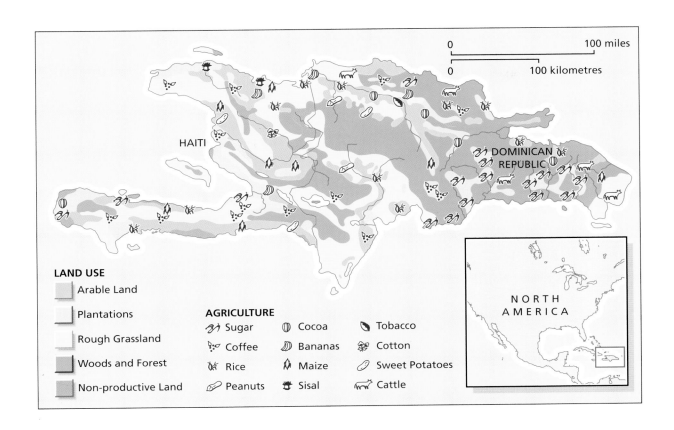

LAND USE		AGRICULTURE		
Arable Land		Sugar	Cocoa	Tobacco
Plantations		Coffee	Bananas	Cotton
Rough Grassland		Rice	Maize	Sweet Potatoes
Woods and Forest		Peanuts	Sisal	Cattle
Non-productive Land				

Changes in island life

Humans on the islands have changed the life around them. Sometimes they have cleared away the natural plants so that they can grow crops. Cutting down too many trees has led to bare islands, with a poor **habitat** for wildlife and difficult conditions for farming.

Powerful nations, such as France, have used islands to test atomic weapons. They have set up military bases on islands too. On the island of Guam, the USA has left behind unwanted tanks and other military equipment.

Hispaniola is often hit by hurricanes. These can cause landslides on the hillsides, which destroy crops in the fields below. Soil erosion is also a big problem.

Looking to the future

The swelling seas

The future of our islands is uncertain. Rising sea levels are caused by the Earth becoming warmer. Some scientists think that temperatures are changing because of the Earth's natural **climatic** cycle. Others think that the Sun is throwing out searing-hot flares.

Many believe that the sun is stronger because the protective layer of ozone gases around the Earth is getting thinner. This could be caused by too many harmful CFC gases being released into the **atmosphere**. Trees absorb these and other carbon gases. But if

The Polynesian Islanders are taking care of their future. Polynesian **atolls** are famous for their black pearls. Thirty years ago, the island of Manihi developed black pearl farms so that more pearls could be sold. They have also encouraged tourists to visit the farms. Manihi has one of the best diving centres in Polynesia too.

we keep cutting them down we are harming the atmosphere even more. Replanting forests, **recycling** wood products and finding ways of using less energy are possible answers.

Support for our islands

Many islands are home to peoples with unique skills, traditions and customs. Island **habitats** have special **species** of plants and creatures too. How can they be protected?

We need to make sure that tourism does not destroy life on the islands. In the Galapagos Islands, only a small number of tourists can visit at any one time. This has protected both islanders and natural island habitats.

Many peoples on small islands depend on tourism and selling just one or two kinds of crop to other parts of the world. These can easily be destroyed by **hurricanes**. Many islands are trying to develop other industries. Richer nations could perhaps support them more in this. They could also pay fairer prices for island crops, **minerals** and goods.

The Hebrides are islands off the north-west coast of Great Britain. Many people make their living by raising sheep and making woollen cloth and clothes. But there is not enough land for everyone. Some people are using computers to set up small computer software or communications businesses. Computers allow people to work away from cities and towns. But others have to leave their islands and find work in other countries, as they have done for hundreds of years.

Island facts

The top ten islands

These are the ten largest islands in the world. Half of them lie in the far northern part of the world.

	Location	Area
Greenland	North Atlantic Ocean	2,175,600 sq km (839,800 sq mi)
New Guinea	western Pacific Ocean	821,030 sq km (317,000 sq mi)
Borneo	western Pacific Ocean	744,360 sq km (287,400 sq mi)
Madagascar	western Indian Ocean	587,040 sq km (226,660 sq mi)
Baffin Island	North Atlantic Ocean	508,000 sq km (196,100 sq mi)
Sumatra	eastern Indian Ocean	473,600 sq km (182,860 sq mi)
Honshu	western Pacific Ocean	230,500 sq km (88,980 sq mi)
Great Britain	North Atlantic Ocean	229,880 sq km (88,700 sq mi)
Victoria	North Atlantic Ocean	212,200 sq km (81,900 sq mi)
Ellesmere	North Atlantic Ocean	212,000 sq km (81,800 sq mi)

Islands and evolution

Did you know that a great scientific discovery was made on the Galapagos Islands over 160 years ago? It was here that the naturalist Charles Darwin, worked out that animals and plants **evolve** separately over time and in different **habitats**.

The coast of Norway is protected all around by a line of thousands of islands called the Skerry Guard. The islands protect the shores from strong waves, wind and **erosion**.

In 1963 a new volcanic island erupted off the south-west coast of Iceland. After three years the volcano started to calm down. It is now nearly 200 metres high and many plants and animals already thrive there. The island was named Surtsey — the name of the ancient Icelandic god of fire.

Glossary

algae simple form of plant life, ranging from a single cell to a huge seaweed

amphibian animal, with a backbone, that develops in water and can stay in the water for long periods, but can also live on land. Frogs, toads and newts are amphibians.

archipelago group of islands. The islands in an archipelago have usually all been formed in the same way.

atmosphere layers of gases that surround the Earth

atoll ring-shaped island made of coral

caribou North American reindeer

climate rainfall, temperature and winds that normally affect a large area

continent the world's largest land masses. Continents are usually divided into many countries.

coral hard rock made of the shells of tiny dead sea creatures, cemented together with limestone made by the creatures themselves

current strong surge of water that flows constantly in one direction in an ocean

delta where a river meets the sea, and the water spreads out into many smaller streams

erosion wearing away of rocks and soil by wind, water, ice or acid

evolve when, over a very long time, creatures and plants develop features and habits which help them to survive well in their environment

fertile rich soil in which crops can grow easily. If you fertilize something you make it fertile.

fjord deep, narrow inlet of the sea with very steep sides

fungus simple plant, such as a mould, mushroom or toadstool

geyser spring that shoots hot water and steam into the air – heated by underground volcanic activity

glacier thick mass of ice formed from compressed snow. Glaciers flow downhill.

habitat place where a plant or animal usually grows or lives

hurricane wind that blows faster than 120 kilometres (75 miles) per hour. Hurricanes can uproot trees and damage buildings.

Ice Age time when snow and ice covered much of the Earth

ice cap large area of ice that mostly covers the land all year round

isolation alone – cut off from the rest of the world

lichen not a true plant – a mixture of a fungus and algae

mammal animal that feeds its young with its own milk

mantle layer of hot, molten rock on which the Earth's crust sits

migrate move from one country or continent to another

mineral substance that is formed naturally in rocks and earth, such as coal, tin or salt

monsoon winds warm, wet summer winds that blow around the Indian sub-continent and the western Pacific Ocean – in winter they are cool and dry

peninsula narrow tongue of land jutting out into the sea

plateau area of high, flat ground, often lying between mountains

polar region area around the North and South Poles

pollinate transfer pollen from one plant to another in order to create fruit and new plants

primate group of creatures including monkeys, lemurs, apes and humans

prow front, or bow, of a ship or boat

recycle reuse

reptile cold-blooded, egg-laying animal with a spine and a scaly skin, such as a crocodile

scour rub hard against something, wearing it away

silt fine particles of eroded rock and soil that can settle in lakes and rivers, sometimes blocking the movement of water

species one of the groups used for classifying animals. The members of each species are very similar.

tectonic plate area of the Earth's crust separated from other plates by deep cracks. Earthquakes, volcanic activity and the forming of mountains take place at the junctions between these plates.

Tropics the region between the Tropic of Cancer and the Tropic of Capricorn. These are two imaginary lines drawn around the Earth, above and below the Equator.

tsunami huge, destructive wave caused by an undersea earthquake

turpentine oil used for cleaning and thinning oil paints

vegetation the plants that grow in a certain area

water vapour water that has been heated so much that it forms a gas which is held in the air – drops of water form again when the vapour is cooled. There is always water vapour present in the air.

Index